U.S. BATTLESHIPS
in action
Part 1

by Robert C. Stern

illustrated by Don Greer

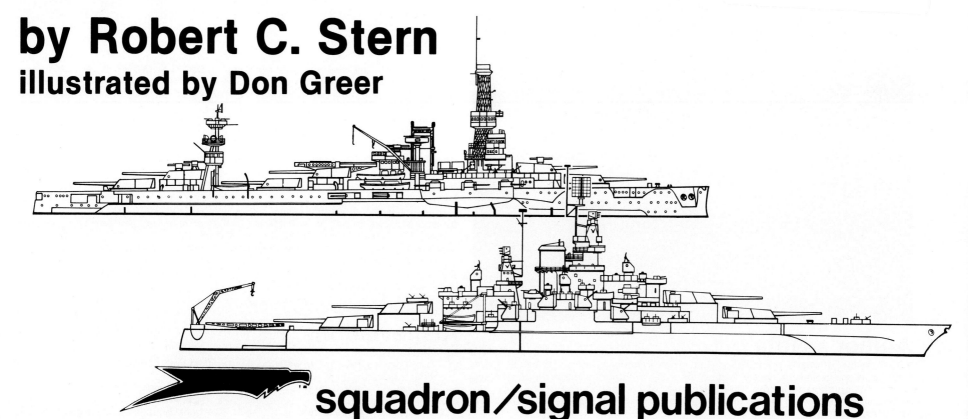

squadron/signal publications

COPYRIGHT © 1980 SQUADRON/SIGNAL PUBLICATIONS, INC.
1115 CROWLEY DRIVE, CARROLLTON, TEXAS 75011-5010
All rights reserved. No part of this publication may be reproduced, stored in a retrieval system or transmitted in any form by any means electrical, mechanical or otherwise, without written permission of the publisher.

ISBN 0-89747-107-5

If you have any photographs of the aircraft, armor, soldiers or ships of any nation, particularly wartime snapshots, why not share them with us and help make Squadron/Signal's books all the more interesting and complete in the future. Any photograph sent to us will be copied and the original returned. The donor will be fully credited for any photos used. Please send them to: Squadron/Signal Publications, Inc., 1115 Crowley Dr., Carrollton, TX 75011-5010.

Acknowledgements

I very much want to thank those people who have helped put this book together. Bob Cressman, as usual, supplied some essential photos and answered numerous questions, Dana Bell pointed out the excellent between-wars shots in the USAF collection and Chuck Haberlein displayed his legendary patience and helpfulness during my semi-annual invasions of the Naval Historical Center. While these and others have helped, any errors of omission or commission in this work are entirely my own.

Nevada, along with other survivors of the holocaust at Pearl Harbor, hands a bit of it back off the coast of Iwo Jima, February 1945. The camouflage scheme is an example (6b) of the relatively rare Ms. 31a, which substituted Navy Blue for Dull Black as the darkest color.

Camouflage

Measure 1 - Dark Gray System (authorized Jan '41) Dark Gray (5-D) on all vertical surfaces up to the level of the funnel tops, Light Gray (5-L) on all higher vertical surfaces.

Measure 5 - Painted Bow Wave (Jan '41) upper edge of false wave in white, filled in with Dark Gray (5-D).

Measure 11 - Sea Blue System (Sept '41) Sea Blue (5-S) on all vertical surfaces.

Measure 12 - Graded System (Sept '41) Sea Blue (5-S) up to level of main deck, Ocean Gray (5-O) up to level of top of superstructure, Haze Gray (5-H) above.

Measure 12 Mod. - The same colors as above but with wavy edge between colors and 'splotching' for a more blended effect.

Measure 21 - Navy Blue System (June '42) Navy Blue (5-N) on all vertical surfaces.

Measure 22 - Graded System (June '42) Navy Blue (5-N) up to level of lowest point of main deck, upper edge horizontal, remaining vertical surfaces Haze Gray (5-H).

Measure 31 - Dark Pattern System (March '43) disruptive pattern generally employing Dull Black (Bk), Ocean Gray (5-O) and Haze Gray (5-H)

Measure 32 - Medium Pattern System (March '43) disruptive pattern generally employing Dull Black (Bk), Ocean Gray (5-O) and Light Gray (5-L).

Measure 33 - Light Pattern System (March '43) disruptive pattern generally employing Navy Blue (5-N), Haze Gray (5-H) and Pale Gray (5-P).

Note: There also existed a more rarely used Measure 31a which substituted Navy Blue (5-N) for Dull Black (Bk), giving a scheme similar in overall effect to Ms. 32.

Author's Note

This is the first of what is intended to be a two part **in Action** series on US battleships in WWII. This one deals with the 'old' battleships, those constructed before the Washington Treaty. The future part will cover the three classes of 'fast' battleships. If any reader has information, anecdotes or photos of these ships, and wishes to share, it will be highly appreciated.

Abbreviations

AA - Anti-Aircraft.
DP - Dual Purpose, referring to weapons used against both surface and air targets and to their directors.
MCG - Medium Caliber Gun.
MG - Machine Gun.
VTE - Vertical Triple Expansion.
FSG - Fire Support Group.
TF - Task Force.
TG - Task Group, a sub-division of TF.

Battleships at work! Rear Adm. Weyler's TG 77.2, Unit 1, supports Operation 'Mike I', the landings at Lingayen Gulf, 8 January 1945. Mississippi leads West Virginia and New Mexico along the coast of Luzon, providing fire support to the GIs who had landed five days earlier. The two lead ships are in Ms. 32 Medium Pattern camouflage (Mississippi - Ms. 32/6d, West Virginia - Ms. 32/7d). New Mexico is in Ms. 21 overall Navy Blue, an effective anti-air scheme that was becoming increasingly popular again with the appearance of kamikazes off the Philippines. (USN-NARS)

Introduction: The Battleship War

Old, slow and ungainly, they had little 'glamor'. They escorted convoys and bombarded shores, leaving the headline-making battles for their newer sisters. So, at least, it seemed from reading those headlines. Yet the reality was quite different. That they were old was undeniable. The newest of them was already 18 years old when WW II broke out, the oldest was an antique 29. A few of them had been re-engined, giving them a top speed of 23kt. Most were still slogging along on their original power-plant and would be lucky to ever see 20kt again. They were far from beautiful, most having been refitted and rebuilt so many times that any notion of the original plan had long since disappeared. Yet, somehow, they managed to be at the right place at the right time, when history hung in the balance. They were at Pearl Harbor on a quiet Sunday morning when America suddenly found itself at war. They were at Surigao Straits, winning the last battleship surface engagement in history, while their newer sisters were off chasing decoys. And, fittingly, they were off the coast of Japan at the end of that great struggle, four long years after it began. While a case can be made that the US Navy's fast battleships were underemployed and misused, no such criticism can be made about the older battleships. They gave all that was asked and more, from the first shot of the war to the last.

The battleships that floated in a double line off Ford Island on that fateful Sunday morning in 1941 were the proud descendants of a long tradition of American battleship design. A claim can be made that **USS Monitor** which fought the Confederate ironclad **CSS Virginia** off Hampton, Virginia, in 1862 was the grandfather of all battleships. Having started the progression, the US Navy soon lagged behind the rest of the world. Satisfied with its monitors, the Navy didn't commission its first battleship until 1895. The 2nd class battleships **Texas** and **Maine** (which sank in Havana harbor) and the 1st class battleship **Indiana** (BB1) were placed in service that year. By the time that war broke out with Spain in 1898, there were four 1st class battleships with the fleet. As individual ships, they were very much up to world standards, mounting four 12" or 13" guns in twin turrets fore and aft with eight 8" guns in twin turrets at the corners. They were better armored and slower than their European contemporaries, a tradition that would be maintained until 1940.

The naval victories at Manila and Santiago were satisfying but also pointed out the numerical weakness of the US Navy relative to other powers, a weakness which the US Navy set out to solve. By the time, Teddy Roosevelt needed an instrument to project US power and prestige around the world, he had it in the 'Great White Fleet', a squadron of 16 battleships that circled the globe between 1907 and 1910. Yet this laboriously constructed fleet was effectively obsolete before it left home. The launching of the **HMS Dreadnought** in 1906 revolutionized battleship design. Mounting ten 12" guns without an intermediate battery, turbine-driven to the then-unheard-of speed of 22kt, **Dreadnought** could choose its battles through superior speed and win them with superior firepower. Her design had originally been established in 1904 and work would have proceded at normal pace except that on 3 March 1905, the US Congress authorized funds for two further battleships, **South Carolina** (BB26) and **Michigan** (BB27). Designed totally independently of **Dreadnought**, these new American battleships were to mount eight 12" guns in superfiring twin turrets fore and aft (a far superior design than that adopted by the Royal Navy - they achieved the same broadside as **Dreadnought** with one fewer turret). Only in the retention of VTE engines, and an 18kt top speed, did **South Carolina** fall behind. The British raced to get **Dreadnought** into the water first, completing her only 18 months after she was laid down. Work on **South Carolina** progressed much more slowly, she wasn't completed until 1910.

The battleline - in spite of the fiasco at Jutland - still, in 1930 represented a nation's might at sea. Here at the Fleet Review for President Hoover off the Virginia Capes, a line of 10 battleships turns in rather ragged unison. Those identifiable are, from the front: Florida, Utah, New Mexico, Nevada **and** Oklahoma. **(NHC)**

Nevertheless, when Britain asked for American assistance after the US entry into WW I, five American 'Dreadnoughts' were available to form the 6th Battle Squadron at Scapa Flow (**Delaware**, BB28; **Florida**, BB30; **Wyoming**, BB32; **New York**, BB34 and **Texas**, BB35 - later **Arkansas**, BB33, replaced **Delaware**). The US Navy also maintained three battleships at Bantry Bay, Ireland for troop convoy escort duty (**Utah**, BB31; **Nevada**, BB36 and **Oklahoma**, BB37). These last two, and **Pennsylvania** (BB38) and **Arizona** (BB39), represented ships that were as well armed and armored as any in the world and only the five British **Queen Elizabeths** were significantly faster.

The British gratitude for the quality and quantity of the American assistance was tinged with fear. Once already in the young century, the Royal Navy had participated in a mad arms race, with the Germans. Now, at a time when the world's economies were badly strained, it appeared that the potential existed for another. On paper, the Royal Navy appeared to have ended the war as, by far, the world's largest fleet, with 33 battleships and 9 battlecruisers. The next largest, the US, had 17 battleships with an additional five under construction with seven more battleships and six huge new battlecruisers funded but not yet begun. The paper figures were illusory, however. The Royal Navy's supremacy, if it existed at all, was much less than the numbers would imply. Many of the Royal Navy's capital ships were old before their time, built hastily to already obsolete designs. Their battlecruisers, as had been forcibly shown at Jutland, were virtually useless. Many of their battleships were slated for immediate scrapping with more to follow. Only one capital ship, **HMS Hood**, was then under construction in British yards and Britain's economy wouldn't allow the kind of ambitious building program that the US was pursuing. The only way to avoid inferiority to the US, or even Japan, which had launched its '8-8' plan in 1916, was to negotiate equality. There was simply no way that war-ravaged Britain could compete with the still fresh economies of the US or Japan.

The invitations finally went out from President Harding in November 1921, after two years of preliminary discussions, for a conference to put the final touches on a Naval Arms Limitation Treaty. All parties were already in basic, if in some cases reluctant, agreement. Capital ship strengths were to be limited in the following proportions:

Great Britain: 5	Italy: 1.75
United States: 5	France: 1.75
Japan: 3	

The British, obviously, came out the best from this. They were allowed 22 capital ships, with permission to build two new ones (**Nelson** & **Rodney**) at the sacrifice of four old. Their only real 'sacrifice' was the scrapping of 11 old Dreadnoughts, already so scheduled, and the abandonment of plans for 12 ships that, without the treaty, probably never would have been built anyway. The US Navy, on the other hand, had to give up 13 of the 15 capital ships then under construction, while four of the earliest Dreadnoughts (**South Carolina, Michigan, North Dakota** and **Delaware**) were to be scrapped. Only the second and third of the 16" **Maryland** class battleships (**Colorado**, BB45 and **West Virginia**, BB48) could be completed. The fourth in that class (**Washington**, BB47), the six **South Dakotas** and the six **Lexington** class battlecruisers were all to be broken up. (A further, grudging concession allowed the US to convert the incomplete battlecruisers **Lexington** and **Saratoga**, and the Japanese likewise **Akagi** and **Amagi**, into aircraft carriers.)

The US Navy was now limited to 18 capital ships. It was essentially these ships that would bear the brunt of fighting in WW II, that would be the target of the Japanese attack at Pearl Harbor. One, **Florida** (BB30), would be scrapped in 1932. Two more, **Utah** (BB31) and **Wyoming** (BB32), would be converted to gunnery training ships, AG16 and AG17 respectively. The remaining 15 were the 'old' battleships that are the subject of this work. They went through varying degrees of modernization during the 1930s. Most had their characteristic cage masts replaced with tripods, their MCG reduced, searchlights and a few AA mounts added. Until the laying down of **North Carolina** (BB55), in 1937, these 15 represented the traditional naval might of the US.

The US Navy maintained Atlantic and Pacific Fleets. As there was no serious, 'traditional' naval threat in the Atlantic that the Royal Navy couldn't handle, and as the Japanese were daily growing stronger and more expansionist, the Pacific Fleet was by far the larger and more important of the two. By 1941, as relations with the Japanese had cooled, that fleet had been advanced from San Diego to Pearl Harbor. Only the creation of the 'Neutrality Patrol' to support the British convoy system in the Atlantic siphoned off some of that strength. From the normal Pacific Fleet complement of 12 battleships, the three **New Mexicos** were transferred to Iceland in June 1941. One further ship, **Colorado**, was at Bremerton undergoing major refit. When the Japanese struck on 7 December, eight battleships, plus **Utah,** were in harbor.

Seven of those eight were berthed on 'Battleship Row' off the southeast shore of Ford Island. **California** was berthed alone at quay F-3, at the southern end of the line. The oiler **Neosho** was at the Gasoline Wharf, next in line. **Maryland** was inboard at F-5, **Oklahoma** outboard of her. **Tennessee** was inboard at F-6, **West Virginia** outboard. **Arizona** was at F-7 with the fleet repair ship **Vestal** alongside. **Nevada** was alone at F-8, at the northern end of the line. **Pennsylvania** was in Drydock #1, along with the destroyers **Cassin** and **Downes**. **Utah** was berthed across Ford Island, in a space normally reserved for aircraft carriers. It was just before 0800 on a peaceful Sunday morning. The crews were up and beginning to muster for morning colors and church service. The surprise could hardly have been more complete.

The Japanese came in two waves, the first of 189 aircraft, the second of 171, but the attack

During the '20s and '30s, most of the 'old' battleships were modernized. Cage masts were replaced by tripods, catapults were fitted, boilers replaced with high pressure models, torpedo bulges added and MCG resited. Here Idaho, **pre-modernization, leads** Texas, **post-modernization, c. 1930. (NHC)**

In a little known foretaste of war, US battleships participated in the 'Neutrality Patrol', operating out of Maine and Iceland, taking over part of the hard-pressed Royal Navy's Western Atlantic duties. The US Navy was at war with Germany in everything but name. Taking it wet, Mississippi negotiates typical North Atlantic weather west of Iceland, 26 September 1941. She wears a very early example of Ms.12 Graded System. (USN-NARS)

tacks appeared continuous to those on the receiving end. Of the first wave, only the 90 Kates (B5N2 Type 97 carrier attack bombers) were specifically assigned naval targets. 40 were armed with torpedoes, the remainder with converted 14" AP shells and 250kg GP bombs for horizontal bombing. All were targeted against 'Battleship Row' or the aircraft carriers, which, in the event, were out of the harbor that morning. Of the second wave, 81 Vals (D3A1 Type 99 carrier bombers) were armed with 250kg bombs for use against the carriers and heavy cruisers. Many of these attacked 'Battleship Row' when they couldn't find their assigned targets.

The attack on 'Battleship Row' began at 0757, when the first torpedo-armed Kates began their run-in from the southeast. Within minutes, **California, West Virginia, Oklahoma, Arizona** and **Nevada** had been holed by the devastatingly effective Japanese torpedoes. **Utah**, despite the fact the she was an unarmed target ship, was mistaken for an aircraft carrier (because of her berth and because her superstructure was partially planked over) and also attacked by torpedo. Because of the shallowness of the harbor and the short space available for a run-in, the Navy had considered torpedo attack impossible within Pearl Harbor. The Japanese, by dint of innovative thinking and rigorous training, had obviously solved the problems. It was the 40 torpedo-armed Kates which did the damage at Pearl Harbor. The horizontal and dive bombers did add considerable weight, but were, of themselves, not crucial. No battleship was seriously damaged by bombs which had not already been wounded by torpedoes.

Admiral Pye's Battle Force was devastated. His flagship, **California** was struck by two torpedoes on her portside, one aft, one near B turret, and one bomb hit starboard amidships. Her torpedo bulkheads remained basically intact, but the effect of the two hits and counterflooding to prevent capsizing, led to her gradual settling. Steps were begun to bring her into watertight condition, which might have prevented her eventual sinking, when oil fires floating down the current forced her abandonment at 1100. By the time she was remanned that afternoon, it was too late. Still, it took three days for her to finally settle on the bottom with a 6½ degree port list. **Oklahoma** sank very quickly. Outboard at position F-5, she took four torpedoes in rapid succession on her portside. Within minutes she turned turtle before any effective correction could be attempted. **Maryland** was inboard of her, as a result taking no torpedoes. Two AP bombs struck, one in the forecastle doing little damage, one underwater at the bow causing considerable flooding. A patch was fitted over the hole in time for her to steam for Puget Sound on 20 December. **West Virginia** was outboard at the next position. Hit rapidly by six or seven torpedoes, she took on an immediate, serious port list. Only the initiative of a pair of seamen in Repair III who, without orders, began counterflooding, kept her from capsizing like **Oklahoma**. She was also struck by two bombs which caused only minor damage. Virtually her entire portside, however, had been opened up by torpedoes, and her sinking was inevitable. Only the timely counterflooding and the fact that she was jammed up against **Tennessee** allowed her to sink in an upright position. At the inboard berth, **Tennessee**, like **Maryland**, was protected from torpedo damage. She was struck by two bombs which caused minor damage. Much more serious was the burning oil leaking from **Arizona** and **West Virginia**. The aft section of **Tennessee** was essentially gutted by the fires, which were sufficiently intense to warp her sternplates and pop many hull rivets. Precautionary flooding of all magazines prevented much more serious damage. Jammed against the quay at F-6 by the sunken **West Virginia**, she could only be extricated by dynamiting the concrete structure on 16 December. On 20 December, she accompanied **Maryland** and **Pennsylvania** to Bremerton.

Arizona was the most seriously devastated of the battleships, the only one to receive major bomb damage. She was at the inboard berth, next up from **Tennessee,** but that position offered no protection. Outboard of her was **Vestal**, which had much shallower draft. One torpedo and at least eight bombs struck **Arizona,** one of those bombs dooming her. Penetrating near A turret, the bomb set off the forward magazines, breaking her in half in front of the bridge. Nearly half of the 2300 dead that Sunday were **Arizona** crewmen. **Nevada** had the most interesting time of them all. Unique among the battleships, **Nevada** was able to get underway, as the prewar doctrine for such situations demanded. Hit by one torpedo in the port bow during the initial attack, counterflooding and a rapidly achieved state of watertight integrity brought the 5 degree list under control. Her machinery was ordered on line as soon as General Quarters was sounded at 0801. Being upcurrent from **Arizona**, she was unaffected by the devastating oil fires, and having no ship outboard, she was free to move when Engineering signalled steam up at 0840. It was the intention of the senior officer onboard, LCdr Thomas, to take her through the Entrance Channel to open water. Backing out of her berth, she swung into the current and headed

Despite War Warnings, no-one expected the Japanese to attack Pearl Harbor. The surprise was complete, the devastation of the battleline was nearly so. (Above) Very early in the attack, but already serious damage has been done. From the left, the battleships are: Nevada, Arizona **with the repair ship** Vestal **outboard,** Tennessee **with** West Virginia **outboard,** Maryland **with** Oklahoma **outboard, the oiler** Neosho **and** California. **The concussion waves are visible from the first torpedo hits, a plume of water rising from** West Virginia's **side. Tracks in the East Loch indicate that more 'fish' are coming. (USN-NHC) (Right) The attack is at its height. Rescue crews, some still in their skivvies, are beginning to react.** The pall of black smoke from Arizona dominates the scene. From the left: California **down by the bow and listing,** Maryland **relatively undamaged,** West Virginia **settled on the bottom and** Oklahoma **capsized. (USN-NARS) (Below) The awful contemplation of the damage, 8 December 1941. Looking over the capsized** Oklahoma at Maryland. **(USN-NARS)**

down 'Battleship Row'. When she was abreast of 10-10 Dock at 0850, she got caught by the Vals of the second wave. Being the only capital ship underway, she was singled out for special treatment. Within a minute, she had been hit by at least five bombs, two in the forecastle, one penetrating the gasoline storage causing a massive fire, another passing through the foremast and exploding at the base of the funnel, another exploding over the crew's galley. The resulting fires threatened the forward magazine which had to be flooded. The fires around the foremast burned out the bridge and cut off all ventilation to the machinery spaces which had to be abandoned. Rear Adm. Furlong, Battle Fleet minecraft commander on his flagship **Oglala,** itself sinking alongside 10-10 Dock, saw **Nevada** in obvious distress, losing way and down by the head. He ordered her to make for the Middle Loch of the harbor rather than risk sinking in the Entrance Channel. She never even made it that far. Shortly after 0900, she beached bow first near the burning **Shaw** at the Navy Yard. Furlong feared that she might swing with the current, blocking the East Loch. He ordered a pair of tugs to pull her free, pushing her across the channel, eventually grounding her stern first near Waipio Point at 1030. There she gradually settled, coming to rest on the bottom Monday morning.

The other battleship at Pearl Harbor that day, **Pennsylvania,** received the least damage. In Drydock #1, she was free from any threat of torpedoing. She was, in fact, hit by only one 250kg bomb which destroyed a 5" mount. The gun was replaced by one from **West Virginia,** the drydock was reflooded on 11 December and she was moved to the Navy Yard the next day. **Utah,** in spite of its disarmed state, was mistaken for an aircraft carrier and brought under attack. Hit by two torpedoes and with only a skeleton crew aboard, she very rapidly listed, coming to rest on the bottom in an inverted position.

The devastation had been total. All eight had been damaged or sunk. Three would be

The sad task of raising and repairing all the victims of the Pearl Harbor attack took nearly three years. Here California is seen in drydock, 9 April 1942, before the water was pumped out. Permanent patches were fitted and she was made ready to steam for Bremerton in early June. The staining caused by her four month immersion reaches up to well above Upper Deck level. (USN-NARS)

Approaching the coast of Axis-held North Africa, Texas escorts the 'Torch' invasion fleet. She then assisted the landings, bombarding Port Lyautey on 8 November 1942. (USN-NARS)

One by one the battleships sunk at Pearl Harbor came back, filling out the battleline and allowing those which had filled the gap a chance to refit. Tennessee returned from her rebuild in time for Operation 'Cottage', the rather comic invasion of Kiska. She is seen here immediately after at Adak, 12 August 1943. (USN-NARS)

After a stint as a radar training ship in early 1942, New York returned to active duty in the Atlantic, participating in the destruction of Jean Bart at Casablanca in November of that year. After more escort duty, she again became a training ship in mid-1943. Seen here in the Chesapeake Bay in early 1944, she shows off the multitude of light and medium AA that have been fitted, at this stage still without director control. (USN-NARS)

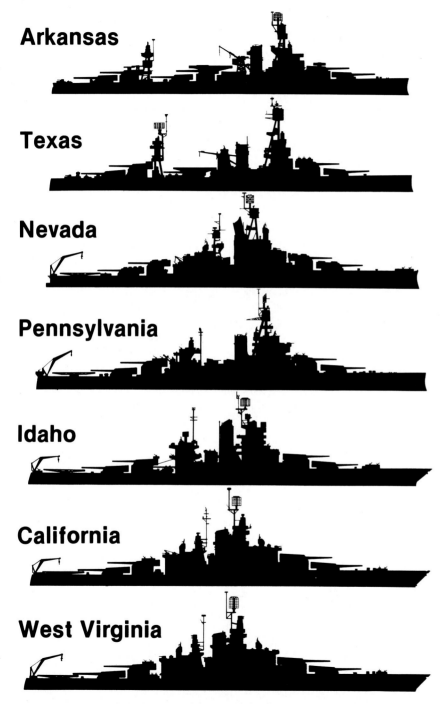

back in service within months and three more would be laboriously raised and rebuilt. The most damaged of those, **West Virginia,** would be out of action for nearly three years. Two more, plus **Utah,** would never be repaired. **Arizona** remains in position today, a memorial to the Navy's dead that Sunday morning. Much has been said about the warning the Navy had of an impending Japanese attack. War Warnings were sent out weeks before 7 December but, when no attack followed, vigilance slackened. Anyway, no-one seriously thought Pearl Harbor would be the Japanese target. Maybe Malaya, maybe Indonesia, almost certainly the Philippines but never Pearl Harbor. The slackened watch and the false confidence that Pearl Harbor was impractical to attack brought about conditions that guaranteed the effectiveness of that attack. Despite at least two War Warnings, 'Battleship Row' was on a peacetime footing that morning. All ships were in 'X' watertight condition, meaning all bulkheads open. Of the ships that sank, only **Nevada** stayed afloat long enough to achieve 'Z' condition watertight integrity. Furthermore, all ships had their fuel bunkers topped off in preparation for a week of maneuvers, feeding the massive oil fires that day. The only ammo available to most AA guns had been the few rounds in the ready-use lockers, which were in many cases locked. On most ships the rapid loss of power meant that ammo lifts from the magazines were rendered useless. Had an effective AA barrage been maintained that morning, some, at least, of the disaster might have been avoided.

On 8 December 1941, the only battleships available to the US Navy were those that had not been at Pearl Harbor the day before. The three **New Mexicos** (which had been joined as TF 1 on 19 July) were in the Atlantic on Neutrality Patrol. **New Mexico** was at Casco Bay, Maine. **Idaho** and **Mississippi** were at Reykjavik. **Arkansas** and **Texas** were with **New Mexico** at Casco Bay and **New York** and **Wyoming** were at Norfolk, Virginia. The only other battleship was **Colorado,** which was undergoing modernization at Bremerton. TF 1 was immediately ordered back to the Pacific, forming up again at San Francisco on 31 January 1942. There they began training and patrolling, escorting convoys to and from Hawaii and standing watch off the West Coast. The caution proved to be unnecessary, but as long as Pearl Harbor remained unusable as a base, and since it would be May before another battleship would be available, perhaps caution was wise. Until May, the three battleships of TF 1, along with Nimitz' three carrier groups which were spending most of their time in the South Pacific, represented the only naval defense for the Continental US. At least in San Francisco they were visible, calming down a nervous population.

1942 was a year of relative inactivity for the 'old' battleships. As they repaired and regrouped, and acted as backstop for the Pacific Fleet, they missed many of the decisive engagements of the war. The battles of Coral Sea and Midway, the Guadalcanal campaign of August 1942 to February 1943 with its string of associated naval encounters, took place without the aid of the 'old' battleships. Gradually the ships less seriously damaged at Pearl Harbor came back on line. **Tennessee,** after hasty and incomplete repairs, joined TF 1 in May. **Colorado,** having had her modernization cut short, and **Maryland** rejoined in time to be detached as a last line of defense off Oahu during the battle of Midway. The two then transferred first to Fiji and then to Noumea, New Caledonia, escorting convoys and patrolling backwaters while three of the new 'fast' battleships engaged the Japanese during the latter part of the Guadalcanal campaign. They remained at Noumea until September 1943. **Pennsylvania** joined the rest of the fleet at San Francisco in July 1942. TF 1, now five battleships strong, transferred to Pearl Harbor in August 1942. Later in the year, as the threat to California and Hawaii appeared to wane, **Tennessee** and **Idaho** returned to Bremerton to complete repairs and undergo major refit.

1943 brought the calm to an end. On 11 May, TG 51.1, composed of **Nevada** (just back from the yard), **Pennsylvania** and **Idaho** began shelling Attu in the Aleutians, preparatory to the landings there. That force remained intact until 2 June when **Nevada** was transferred to the Atlantic, **Pennsylvania** went to Mare Island for a much needed modernization and **Idaho** returned to Pearl Harbor. For the invasion of Kiska in August, TG 16.22 was formed under Rear Adm. Griffin. On 27 July, that force, on that day consisting of **Mississippi** and **Idaho,** fought the historic 'Battle of the Pips' 80 miles west of Kiska. Radar was still a new and unreliable tool and the weather in the Aleutians is nearly always atrocious. Visibility that day was normal, near-zero. 518 rounds of 14" shells were fired by the two battleships at a series of radar contacts. Unfortunately, there were no Japanese warships within 200 miles. 19 days later, on 15 August, **Pennsylvania, Idaho** and **Tennessee,** now reformed as TG 16.17 under Rear Adm. Kingman, put their guns to more positive effect, shelling Kiska in support of Operation 'Cottage'. 34,000 troops were landed on an island that was occupied by a handful of stray dogs. The 5100 defenders had been withdrawn under the very nose of the US Navy on 28 July, the day after the 'Battle of the Pips'.

Operation 'Galvanic', the invasion of the Gilberts began on 20 November 1943. The Northern Attack Force (TG 52.2) under Rear Adm. Griffin, composed of **New Mexico, Pennsylvania, Idaho** and **Mississippi,** shelled Makin. The Southern Attack Force (TG 53.4) under Kingman, **Tennessee, Maryland** and **Colorado,** did the same at Tarawa and Abemama, remaining in position until resistance had ended on 28 November. They became a fixed part of Nimitz' Central Pacific strategy. Each invasion target was to be 'softened up' by carrier strikes and a rapid bombardment by the 'fast' battleships. Then came the landing, backed up by the older battleships supplying tactical support, the range of their guns often allowing them to intervene at any point on the island. Unlike the carrier air, this support was always available, day or night, with greater accuracy, at lower cost and risk in sustained quantities. As the Japanese were to find to their distress, time and again, the effect of even a few salvos of 14'' shells was often enough to break up an attack or crumble a defense.

Next came the Marshalls, Operation 'Flintlock'. Griffin's TG that had shelled Makin was rechristened the Southern Attack Force FSG 52.8, now under Rear Adm. Giffen, and sent against Kwajalein. Likewise, the Tarawa force was given a new CO, Rear Adm. Oldendorf, a new name, Northern Attack Force FSG 53.5, and a new target, Roi. Both groups were committed to 'Flintlock' from 31 January to 7 February 1944. The capture of Eniwetok atoll at the western edge of the Marshalls was considered a separate operation, codenamed 'Catchpole'. The FSG for the invasion fleet, TF 51, was composed of **Tennessee, Colorado** and **Maryland,** under Oldendorf. Resistance continued from 17 to 23 February.

A temporary lull in the drive across the Central Pacific followed the capture of the Marshalls. Taking advantage of this, MacArthur 'borrowed' four of the 'old' battleships and Rear Adm. Griffin. **New Mexico, Tennessee, Idaho** and **Mississippi** were used on 20 March 1944 to shell Kavieng as a diversion from the Army's invasion of Emirau Island in the St. Matthias Group of the Bismarck Archipelago. 1300 rounds of 14'' and 5'' ammo were expended on the town.

On the opposite side of the world, the 'old' battleships were back in action again in June. On 6 June, the three active battleships in the Atlantic initiated Operation 'Neptune',

In a typical pose, with landing craft and a troop carrier in the foreground, West Virginia, **the last to return from Pearl Harbor, adds her weight to the battle for Leyte, 24 October 1944. Part of the Fire Support Group of TF 78, she would that evening be diverted to Surigao Strait for the last battleship vs. battleship engagement ever fought.** (USN-NARS)

On the night of 24-25 October, 1944, the six battleships of the combined FSGs of TF 78 and 79, under Rear Adm. Oldendorf, blocked the exit from Surigao Strait in the path of Nishimura's two battleships. Only **Yamashiro** reached the line of American battleships, where she was literally blown to pieces. **West Virginia** is barely visible against the flash of gunfire and smoke of battle, as seen from **Pennsylvania**. (USN-NARS)

the naval component of 'Overlord', the invasion of France. (This wasn't the first 'operational' use of the Atlantic battleship force, **New York** and **Texas** had briefly supported the 'Torch' landings in November 1942.) Force 'A' under Rear Adm. Deyo, including **Nevada**, backed up the landings at Utah Beach. Force 'C', Rear Adm. Bryant, with **Texas** and **Arkansas**, operated at Omaha Beach. All three remained in action until the front moved inland beyond effective range. They again came into action as a group during the assault on Cherbourg. The comments of Army staffers about the effectiveness of the naval bombardment during the Normandy battles are enlightening. The Chief of Staff of the 1st Infantry Div. stated: "...without that (naval) gunfire we could not have crossed the beaches". The enemy as well gave grudging praise. Field Marshall Gerd von Rundstedt stated: "The fire of your battleships was a major factor in hampering our counterattacks."

Operation 'Forager' was the codename for the invasion of the Marianas, as crucial in the Pacific as 'Overlord' was in the ETO. TG 52.17 under Oldendorf was composed of **Tennessee, California,** (just returned from repair and modernization), **Maryland** and **Colorado**. TG 52.10 under Rear Adm. Ainsworth included **Pennsylvania, Idaho** and **New Mexico**. Both groups were used for the pre- and post-invasion bombardment of Saipan between 14 June and 9 July 1944. On 22 June, the first major casualty since Pearl Harbor was suffered, **Maryland** being hit by an aerial torpedo. The damage wasn't critical, but was sufficient to require her return to Pearl Harbor for two months of repairs. On 19 July, the six remaining battleships of the two groups were merged into TG 53.5 for the bombardment of Guam. On 24 July, the target was Tinian. On that day, **Colorado** received 22 hits from coastal batteries, sending her back to Pearl Harbor until November.

The 'old' battleships entered yet another sea as part of Operation 'Dragoon', the invasion of Southern France, beginning 15 August 1944. TF 85, the 'Delta' Force FSG was composed of **Texas** and **Nevada** under Bryant. TF 87, the 'Camel' Force FSG was based on **Arkansas** under Deyo. Both groups participated in the bombardment of Toulon.

Pennsylvania, Tennessee, Maryland, Mississippi and **West Virginia** (the last of the boats 'sunk' at Pearl Harbor to return) formed the TF 31 FSG under Oldendorf for Operation 'Stalemate II', the conquest of the Palaus. Pre-invasion bombardments were carried out against Peleliu and Angur Islands on 13-14 September 1944.

The reconquest of the Philippines was to be the most exciting and time-consuming operation for the battleline. Beginning with the bombardment of the Leyte beaches on 20 October 1944, the 'old' battleships would be involved in the Philippines into February 1945. The most excitement was at the beginning. The initial pre-invasion bombardments went routinely well enough. **Mississippi, Maryland** and **West Virginia** comprised the Northern Attack Force FSG, TF 78 under Rear Adm. Weyler. The FSG for the Southern Attack Force, TF 79, was composed of **Tennessee, California** and **Pennsylvania** under Oldendorf. The routine of coastal bombardment was broken by the Japanese reaction to Leyte landings. While the reaction to the Saipan invasion had been expected and the resulting Battle of the Philippine Sea hadn't involved the 'old' battleships, Japanese response now was

Idaho seen from Nevada, cruises off the shore of Iwo Jima, with Mt. Suribachi in the background, February 1945. By this stage of the war, Ms. 21 was again the standard camouflage.

Off Okinawa, because of their very usefulness for fire support, the 'old' battleships were vulnerable to the retaliation of the kamikazes. (Above Right) Tennessee lays down the last salvos before the landing craft hit the beach. Typical of the modernized battleships, she displays the extra beam that allowed the mounting of twin 5" DP turrets. (USN-NARS) (Below Right) Suddenly on the receiving end, Tennessee took a kamikaze aft, 12 April 1945. In the background, Zellars (DD777) also burns, another kamikaze victim. The damage to Tennessee wasn't serious. (USN-NARS)

something of a surprise and the American counterstrokes were unco-ordinated and at times ill-advised. Ozawa's decoy force was sighted north of Cape Engano and Halsey's carriers, with Lee's 'fast' battleships in tow, set off in pursuit. Therefore when the forces of Nishimura and Shima were sighted heading for the southern entrance to Leyte Gulf on 24 October, the task of guarding the beaches from the Japanese fell to the 'old' battleships. The job of organizing the defense was turned over to Oldendorf who arrayed his forces in depth. As Nishimura's Force C (the 'old' battleships **Fuso** and **Yamashiro,** the cruiser **Mogami** and four destroyers) entered the Mindanao Sea, it encountered the first line of PT boats, but got through unscathed. At 0115 on 25 October, the force entered the nearly 40 mile long Surigao Strait between Leyte and Dinigat at the end of which lay the prize, the US invasion fleet in Leyte Gulf. Between Force C and that prize was one more line of PTs, three groups of destroyers, then cruisers and then the battleline. Nishimura never really had a chance. The first line of destroyers disabled **Fuso**, which sank at 0307. The next line torpedoed three of his four destroyers, sinking two, and put a torpedo into **Yamashiro.** At 0225, she was hit again by the last of the destroyer lines but kept on coming, right at Oldendorf. The cruisers opened fire at 0351, followed by the battleline two minutes later, at a range of 22,800yd. **West Virginia, Tennessee** and **California,** which had the latest Mk 8 (FH) fire control radar, contributed most of the firepower. 225 rounds of 14" & 16" AP were fired in limited salvos to conserve a short supply. **Maryland** ranged on **West Virginia's** salvos and got off 48 rounds. **Mississippi** fired only a single salvo, while **Pennsylvania,** masked by the other ships, didn't fire at all. **Mogami** returned fire until 0355 when, battered, she turned away. **Shigure,** Nishimura's remaining destroyer, threaded her way through the shell splashes, retiring after receiving only one minor hit. **Yamashiro,** with Nishimura onboard, returned fire until almost 0410. Then, afire from end to end, she too turned away to the south. Oldendorf ordered fire ceased at 0409. Incredibly,

Yamashiro was still capable of fighting back. When destroyers closed to finish her off, she put up a sustained medium calibre fire, turned again and increased speed to 15kt, hoping to make good her retirement. A minute later, her fate was sealed by two additional torpedoes. At 0419, she capsized and sank. For the last time, battleship would fight battleship. It is fitting that it should have been Oldendorf's 'old ladies' administering the *coup de grace*.

Shima's cruiser force was trailing Nishimura by about an hour, following the same track through the Mindinao Sea. His lead cruiser brushed against the retiring **Mogami**, which, ablaze from innumerable hits, managed to work up to 18kt, and tag along with Shima as the cruiser force now retired. Despite her efforts, however, her fate was to sink that day. Three of Oldendorf's cruisers caught her again at 0530 and added some more damage. Two PT boats attacked at dawn, slowing her further. Avengers found her at 0810, put a torpedo into her hull and left her for dead. She was now finally helpless, drifting dead in the water. Shima's destroyer **Akebono** took off her crew and sank her with another torpedo. **Shigure**, essentially undamaged, was the only survivor of Nishimura's force, making Brunei on 27 October.

At approximately 0530, Oldendorf led his victorious battleline back into Leyte Gulf for essential replenishment of depleted ammo stocks. Thus his fleet wasn't available to intervene in the battle between Kurita and Taffy 3 which began at 0648 off Samar, nearly 100 miles to the north. Until Kurita arrived unannounced, Admiral Kinkaid, in charge of the invasion fleet, wasn't aware that the San Bernardino Strait, the northern entrance to Leyte Gulf, had been left uncovered by Halsey's withdrawl of Lee's 'fast' battleships. Those battleships were almost 400 miles away, chasing shadows. Oldendorf's replenishment was cut short, but there was no real chance of his being able to intervene. Lee's TF 34 wasn't released by Halsey until 1055. When they arrived off San Bernardino at 0100 on 26th, they had missed Kurita's retreat by three hours. There had been, up to this point, an unspoken rivalry between the 'old' and the 'fast' battleships. The events of 25 October 1944 gave the 'oldies' bragging rights for the duration.

The 'old' battleships stayed in Leyte Gulf long after Kurita's departure. In November, the force was re-organized and trimmed down. TG 77.2 was formed under Rear Adm. Weyler with **Maryland, West Virginia, Colorado** and **New Mexico**. Inevitably the group became a target for the new kamikaze suicide attacks. **Colorado** was struck twice on 27 November without serious damage. Two days later, **Maryland** was hit between A and B turrets, causing serious damage, forcing her return to Pearl Harbor until March 1945. Minus **Maryland**, and with a new name, TG 77.12, and CO, Rear Adm. Ruddock, the three remaining battleships struck at Mindoro and the Western Visayans on 15 December.

The force was totally reconstructed on 3 January 1945 for the invasion of Luzon at Lingayen Gulf, Operation 'Mike I'. TG 77.2, under now Vice Adm. Oldendorf, was composed of **Mississippi, West Virginia** and **New Mexico** in Unit 1 and **California, Pennsylvania** and **Colorado** in Unit 2. Three days later, **New Mexico** and **California** were both struck by kamikazes. The former was hit on the bridge for minimal damage. **California** was more seriously hit, requiring a month and a half at Puget Sound. Three days later, another pair of Oldendorf's battleships were hit. **Mississippi** was hit on the bridge by a kamikaze and **Colorado** was hit by coastal artillery but neither was seriously damaged and neither had to retire.

Because of a continuing need for the presence of Oldendorf's TG at Lingayen, a new force had to be gathered for Operation 'Detachment' at Iwo Jima. Gathering two battleships returning from refit (**Tennessee** and **Idaho**) and four from the Atlantic, no longer needed there (**Nevada, Texas, New York** and **Arkansas**), a force of six was put together as TF 54 under Rear Adm. Rodgers for use on 16 February.

The last big amphibious operation of the war gathered all the 'old' battleships together for the last time, the invasion of Okinawa, Operation 'Iceberg'. Not only was this to be a grueling land campaign for the Army and Marines, one requiring almost daily intervention by the battleship's big guns, but it was to be exceedingly dangerous for those ships as well, which became the target for succeeding waves of kamikazes. The ten battleships that formed TF 54 under Rear Adm. Deyo were broken down into five two-ship sections: Gp. 1 - **Texas** and **Maryland**, Gp. 2 - **Arkansas** and **Colorado**, Gp. 3 - **Tennessee** and **Nevada**, Gp. 4 - **Idaho** and **West Virginia** and Gp. 5 - **New Mexico** and **New York**. They went into action on 26 March 1945 and began to suffer casualties immediately. On the 27th, **Nevada** was hit for minor damage. On 1 April, it was **West Virginia's** turn and on 5 April, **Nevada** was hit again, this time by five shells from a coastal battery. The first serious damage came on 7 April, when **Maryland** was hit on the side of C turret, the bomb penetrating the armor deck. She returned to Puget Sound for repairs, not returning until after the war was over. Both **Tennessee** and **Idaho** received minor damage on 12 April, as did **New York** two days later.

On 18-19 April, the initial storming of the Shuri defense line gave the 'old' battleships one more chance to prove their worth. Again, toward the end of the month, **Mississippi** put 1300 14" shells into Shuri Castle in limited visibility. The newer **North Carolina** and **South Dakota** had tried to reduce Shuri a few days before without success, but **Mississippi** hit it on her first salvo and left it a pile of rubble in a matter of hours. On 12 May, **New Mexico** was hit by a kamikaze, requiring a return to Manila for repairs. On 5 June, **Mississippi** was hit off Ineya Shima. The last battleship to suffer kamikaze damage was **West Virginia** on 17 June.

TF 95 was formed on 16 July under Vice Adm. Oldendorf for operation in the East China Sea. Six 'old' battleships (**Tennessee, Pennsylvania, California, Nevada, Arkansas** and **Texas**) were joined by the new 'battlecruisers' **Alaska** and **Guam**. In their only offensive sortie, **Nevada, Tennessee** and **California** along with the battlecruisers raided the Yangtse Estuary off Shanghai between 26 and 28 July. Few targets remained for the big guns and little was achieved by the raid.

Having found little of interest in the East China Sea, TF 95 was broken up for various other tasks. The only unit to see further action was TG 95.1 composed of **Tennessee** and **Pennsylvania** which were sent to shell the Japanese garrison on Wake. The two ships bombarded the island together on 1 August. Eight days later, **Pennsylvania** repeated the job by herself. On 12 August, a single Japanese aircraft put an aerial torpedo into her stern quarter for serious damage. She was towed into shallow water as a precaution against her sinking, though, in the event, the flooding was brought under control. After emergency repairs, she was towed back to Bremerton for further repair work.

Only one more task remained for the proud 'old ladies' of the US Navy. Entering Tokyo Bay along with Halsey's TF 38, **New Mexico, Mississippi, Idaho, Colorado** and **West Virginia** were there to witness the surrender of the enemy that had left several of their number as smoking wrecks only four years before.

With the war over, the fate of the 'old' battleships was sealed. What the enemy wasn't able to accomplish, age and the parsimony of a peacetime congress achieved with grim swiftness. Of the 14 'old' battleships in service at war's end, four were immediately designated as target ships for the Bikini Island atomic bomb tests of July 1946. Another three were decommissioned in 1946, four more in 1947 and one in 1948. Only **Mississippi** soldiered on, redesignated EAG128 on 15 February 1946, to act as a testbed for anti-aircraft systems. She remained in service in this form until 1956. One by one, the 'old' battleships were struck from the Navy List and sold for scrap between 1947 and 1959. Only two of them now remain. **Arizona** lies still in the mud off Ford Island, still a commissioned warship of the US Navy. **Texas** remains, somewhat worse for the wear, in the waters off Houston, a sad reminder of what can happen to a nation caught unprepared.

The US Navy in WW II was in the enviable position of having enough of the 'old ladies' around that the devastation of Pearl Harbor and the fury of the kamikazes could be survived with ships to spare. Considered by many in 1941 to be too old and slow to be of much use, they proved instead to be invaluable at a myriad of unexpected tasks. The lesson to be drawn from this, that a nation must have hulls in the water to fight a naval war, must not be lost on the current generation of legislators and planners.

New York adds her contribution on the side of the Marines on Iwo Jima, 19 February 1945. The pinpoint accuracy of the 'old' battleship's gunfire could reach anywhere on the island, often proving decisive against the stubborn Japanese resistance. The lack of such fire support could affect the US Navy's ability to repeat such operations today.(USN-NARS)

No longer an active combatant, but a casualty at Pearl Harbor nonetheless, Utah had already been converted to target ships when seen here at San Pedro, 18 April 1935. On 7 December 1941, she had wooden planking over most of her upperworks, causing her to be mistaken for an aircraft carrier. She took a pair of torpedoes and capsized. (USN-NARS)

Wyoming Class

AG17 (ex BB32) Wyoming, BB33 Arkansas

These two represented the last class of 12''-armed US battleships. Following the initial four-turret **South Carolina** class of 1910 and the five-turret **Delaware** and **Florida** classes, the **Wyomings** mounted six twin turrets. (This represented nearly the limit in the multiplication of centerline-mounted twin turrets, only **HMS Agincourt,** with seven, had more.)

Between 1925 and 1928, both ships underwent major refit, having torpedo bulges added. The 12 original mixed-firing boilers were replaced by four oil-fired, allowing the two funnels to be reduced to one. The after cage mast was removed and a short tripod mainmast installed between R and X turrets. On each beam, three of the five 5''/51 casement guns was raised one deck to unarmored positions. A catapult was fitted to Q turret and the appropriate aircraft crane shipped.

Wyoming was demilitarized in 1932, Q, R and X turrets and her main armor belt being removed. At that time she was redesignated AG17. She was employed as a gunnery trainer, unmodified, until 1944. At that time, she finally lost her cage foremast, her bridge was extensively rebuilt and a pole foremast was stepped. Her role was changed to that of an MCG trainer. All 5''/51s and the three remaining 12'' turrets were removed and 14 5''/38 DP guns mounted (5 x 2, 4 x 1).

Arkansas was at Casco Bay, Maine, as part of the Neutrality Patrol, when war broke out. A 1940 refit had increased the maximum elevation of her main guns. In 1942, she underwent another major rebuild. Her cage foremast was replaced by a tripod and her four remaining main deck casement guns were removed. She remained in the Atlantic on convoy escort through 1944. After a refit including enlargement of her bridge, modifications to her mainmast and upgrades to her anti-aircraft and radar fits (her final AA fit was 9 quad 40s and 26 single 20s), **Arkansas** took part in the invasion of Europe and then came to the Pacific for the Iwo Jima and Okinawa operations.

Arkansas was expended as a target at Bikini in 1946. **Wyoming** was stricken in 1947.

Before radar and naval aircraft, when gunfire still ruled the waves, Wyoming represented naval power at its epitome. She is still in the weather-beaten medium gray she wore at Scapa Flow a year earlier. The most noticeable additions, borrowed from the Royal Navy, are the deflection scale painted around X turret and the range clock on the mainmast. The British had found at Jutland that visibility in actual battle was often so bad that only a few ships in a battleline could see the enemy at any one time. The deflection scale and range clock allowed ships which couldn't see to obtain essential firing data from observation of those which could. Wyoming is seen here passing through the Panama Canal, 26 July 1919. With Germany eliminated, Japan was already perceived as the main threat. (USN-NHC)

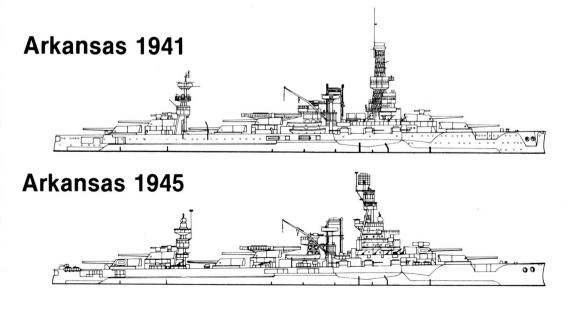

Arkansas 1941

Arkansas 1945

Post-modernization, Arkansas has lost her cage mainmast, October 1931. Thought to be light and strong, cage masts proved to be strongly affected by winds, at least one collapsing in a storm. Curiously, her catapult is being used for boat stowage. Note the proliferation of main armament rangefinders, at least four are visible, two on X turret alone. (USN-NARS)

Largely unchanged, Arkansas is now at war, February 1942. Perhaps the only visible changes are the AA platforms atop her boat cranes and mainmast, SC radar (seen end on) atop her mainmast and her Ms. 11 overall Sea Blue camouflage. (USN-NARS)

By 1 July 1943, Arkansas had gone through another major modernization, losing her cage foremast and gaining more AA and radars (FC and SG on her foretop, SC still on her mainmast). Her primary duty at this time was convoy escort in the Atlantic, to discourage sorties by German raiders. She is in Ms. 22 Graded scheme, the 'standard' Atlantic scheme, effective against detection by submarines. (USN-NARS)

A bit of excitement off Normandy. Arkansas gets a chance to use her guns, 6 June 1944. She was with Force 'C' off Omaha Beach on D-Day. (USN-NARS)

By the time Arkansas reached San Pedro, 1 January 1945, the emphasis of camouflage had changed from anti-submarine to anti-air and the Ms. 31 gave way to Ms. 21. The microwave dishes on the secondary armament directors are plainly visible. (USN-NARS)

Her Atlantic duties done, Arkansas refits at Boston, 5 November 1944. This view shows her with SK, SG and FC on her foremast and SG repeated aft, a common practice because of problems with reliability of SG information. She now at last carries secondary battery control in the form of directors with microwave dishes on her forebridge and mainmast. Two of her main battery rangefinders have been landed. She is wearing a striking Ms. 31a/7b, preparatory to transfer to the Pacific. (USN-NARS)

Wyoming spent the war as a gunnery trainer, until 1944 retaining her main turrets. She carries nearly every AA weapon in the Navy's arsenal, the port and starboard fits varying considerably. Portside she carried assorted open, single 5" mounts. Starboard she had turretted single 5"/38s. At her bridge wings she had different AA directors, a Mk 37 with FD radar starboard and an older, pre-radar director portside. She still carries her range clock on her mainmast, 17 June 1943. (USN-NARS)

AA Drone

An early 1944 refit saw Wyoming's 12" turrets replaced by twin 5"/38s. At least two additional AA directors have been shipped on the centerline between the masts, 18 April 1944. (USN-NARS)

In 1916, the war in Europe looking more and more like it would involved the US, the Navy increased the frequency and difficulty of exercises. Texas (in the foreground), New York (behind) and three other battleships lie at Guantanamo, Cuba, 1916. Both Texas and New York carry an 'E' on their after funnel, indicating excellence in some practiced skill, most likely gunnery. (USN-NHC)

New York Class

BB34 New York, BB35 Texas

Desiring to continue the process of strengthening the armament of each succeding class of battleships without resorting to a seventh twin 12" turret, BuShips was faced with the option of selecting triple turrets or a larger caliber main armament. Convinced by the Royal Navy's decision to mount 13.5" guns in their **Orion** class laid down in 1909, they opted for the latter, choosing to arm the **New Yorks** with fewer (10 in five twin turrets), larger (14"/45) guns.

Anachronistically, these two reverted to VTE reciprocating engines, from the otherwise standard turbine main propulsion. Ostensibly this was because the VTE gave more economical cruise, but equally it represented a streak of hidebound conservatism among US Navy planners.

A 1926-27 rebuild left the **New Yorks** altered much as the **Wyomings** had been. A smaller number of oil-fired boilers replaced the old coal-burners, allowing a reduction from two funnels to one. Similarly, six of the casemented 5"/51s were moved up a deck to make room for a torpedo bulge. A tall tripod foremast and short tower mast replaced the two cage masts. A short tripod mainmast was added between Q and X turrets.

Both **New York** and **Texas** were in the Atlantic at the outbreak of war. **New York** was at Norfolk as a radar trainer, **Texas** was at Casco Bay. Both ships took part in the 'Torch' landings in November 1942. From 1943, **Texas** had a platform for six single 20mm Oerlikons on B turret. In 1944, for the Normandy landings, **Texas** carried an extra tall top on her foremast for monitoring German electronics. **New York** spent this period as a training ship on both coasts. Both met again as part of Adm. Rodgers' TF 54 off Iwo Jima. Final AA fit was 10 quad 40s and 36 (**NY**-30) 20s.

New York became a target at Bikini in July 1946. **Texas** was more fortunate. Decommissioned in 1948, she became a memorial at the San Jacinto Battlefield near Houston, the only 'old' battleship to have survived intact.

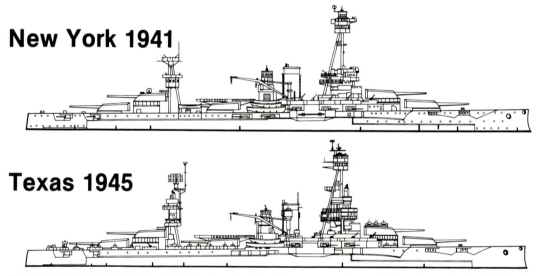

New York 1941

Texas 1945

Having been modernized in the '20s, the New York lost her cage masts and a funnel, and had a citadel raised amidships to carry part of the MCG. Q turret has a catapult. Atop A, B, X and Y turrets are sub-caliber spotting rifles used frequently during gunnery practice to save wear on the main gun tubes. The total AA fit is four 3"/50s each side atop the citadel and a pair of MGs on foremast platforms, Hampton Roads, 17 October 1929. (USN-NHC)

New York in Ms. 12 Mod. Graded System, the so-called 'splotch' scheme, at Norfolk, 1 February 1942. After having been the testbed for the XAF radar in 1938, she now mounts an FC on her foretop and another on the tower aft of her funnel. Her mainmast has been converted into a 'flak tower' with 20mm Oerlikons at two levels. Note the guntubs for her 3"/50s. (USN via Floating Drydock)

Later in 1942, Texas carries a complex Ms. 12 Mod. In the course of the year, she has had her AA multiplied many times over, with two single 20mm next to B turret, a quad 40mm forward of the foremast legs, another next to the funnel, at least five single 20s next to Q turret and another quad 40 and a 3"/50 further aft. More single 20s extend aft of Y turret with a pair of quad 40s at the stern. (USN-NARS)

A clear aerial view of Texas, 15 December 1942, showing graphically the explosion of light and medium AA guns. While largely uncontrolled, the sheer weight of AA fire would be a deterrent. (USN-NARS)

(Above Left) Both ships participated in the 'Torch' landings, both now in Ms. 22. The attempt to follow the horizon with the top edge of the Navy Blue has given New York a curiously swayback appearance, Casablanca, 10 November 1942. (USN-NARS)

Texas in December 1943 showing the continued expansion of the light AA. Platforms with six 20mm mounts now top B and Y turrets, one more quad 40 has been fitted forward on each side as well as numerous 20s, some of the existing mounts being resited. Her mainmast now carries an SK air search and SG surface search radar. (USN-NARS)

After the 'Dragoon' landings in Southern France in August 1944, there wasn't too much more work for a battleship in the Atlantic and Texas transferred to the Pacific in time for Iwo Jima. (Right) Seen on 6 January 1945, Texas wears a fresh Ms. 21 coat of overall Navy Blue. As can be readily seen, this camouflage, when new, could be extremely effective in preventing detection from the air, at least when the ship isn't throwing off a wake of white water. (USN-NARS) (Below) Off Iwo Jima, a month later, the Navy Blue has now aged and faded, as all predominantly blue pigments did, making deception somewhat less effective. The radar suite is the same as in 1943 except that the arrangement of the mainmast has been altered and another SG has been added, on the foretop. On her mainmast and forebridge, Texas now has AA directors with microwave radar. (USN-NARS)

(Below Right) Reprieved at last from training duties, New York rejoined the fleet in 1945. Seen here off Okinawa, 3 March 1945, she is wearing an attractive Ms. 31a/8b. Her radar fit is the same as Texas' but arranged differently. Her AA fit is also similar, mainly lacking the batteries on B and X turrets. (USN-NHC)

Nevada Class

BB36 Nevada, BB37 Oklahoma

This class introduced a new armor scheme to the US Navy and, as a consequence, a new arrangement of the main battery. Experiments with the old target ship **San Marcos** (ex-**Texas**) in 1912 led US designers to conclude that the medium and light armor that was liberally applied to the bow, stern and upperdecks of contemporary battleships had no protective value and was in fact a disadvantage. It could cause the explosion of a shell that might otherwise pass through and it used up displacement that might be better used elsewhere. Obviously, if that armor was eliminated, the weight saved could be used for thickening the main armored citadel, hence the 'all-or-nothing' or 'raft body' principle of protection introduced in the **Nevada** class. The idea was that the smallest possible area enclosing the vitals (boilers, engines, magazines, etc.) would be given the greatest possible protection. The 'raft body' had to be large enough that major flooding fore and aft wouldn't endanger buoyancy, yet small enough that the greatest armor thickness could be obtained for a given weight of armor. Several other expediants were used to save weight or length in the vitals. The **Nevadas** were built with oil-fired boilers, saving significant weight and volume. The only negative side of this was the loss of the protective value of coal bunkers, which partially offset the gain in armor thickness. A further shortening of the citadel was achieved by reducing the number of turrets by one. By adopting a triple-mount at A and Y positions, Q turret could be eliminated without sacrificing firepower. All in all, a more useful and 'graceful' ship was achieved in comparison with the preceding classes. Speed dropped by a half-knot (to 20.5kt) but maximum protection was increased from 12" to 14" and the extent of that maximum thickness greatly increased. **Oklahoma** retained VTE engines, although **Nevada** became the first American battleship to be powered by geared turbines, whose lower ratio cruise settings gave fuel economy similar to that of reciprocating engines.

Both ships were modernized in the late '20s. Boilers were replaced, torpedo bulges were added and horizontal armor was increased. The raised forecastle design allowed the casement MCG to be carried a deck higher than in the previous classes. Nevertheless, experience showed them still to be too wet and they were raised another deck at this time. The most noticeable change was the replacement of the cage masts by tripods and the siting of catapults on X turret and the fantail. The funnel on both ships was raised in 1936.

Both were at Pearl Harbor on 7 December 1941. **Oklahoma** capsized within minutes of the initial attack, **Nevada** was the only battleship at Pearl Harbor to get underway, reaching the Entrance Channel before succumbing to multiple bomb hits, settling in an upright position. **Oklahoma** was intended to be raised but was eventually written off, as much because of her old-fashioned VTE main engines as the seriousness of her damage. **Nevada** was raised and moved to Puget Sound for modernization and repair. Only in the spring of 1943 did she rejoin the fleet. Her large after tripod had been replaced by a stump mast and the foremast reduced in size to improve the arcs of fire for the increased AA armament. The 5"/51 MCGs and 5"/25 AA guns were replaced by eight twin 5"/38 turrets. The X turret catapult was removed and, most noticeably, an angled funnel extension was fitted which gave **Nevada** a unique appearance.

After the Aleutian operation in May 1943, **Nevada** was transferred to the Atlantic to replace **New York**. Having taken part in the Normandy and Southern France invasions, she transferred back to the Pacific in time to participate in operations at Iwo Jima and Okinawa. She ended the war operating in the East China Sea with TF 95.

Oklahoma was righted and raised in 1943, but was never rebuilt, being sold for scrap in 1946. Like her predecessors, **Nevada** had no future in the postwar Navy. Surviving nuclear tests Able and Baker, she was sunk as a gunnery target in July 1948.

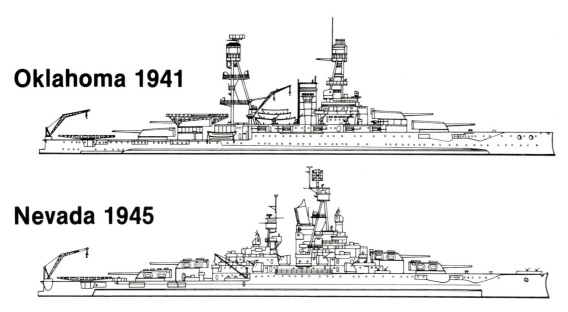

Oklahoma 1941

Nevada 1945

The Nevadas differed from the preceding class externally in the rearrangement of the main armament. By adopting triple turrets in A and Y positions, the same number of guns could be carried in one fewer turrets. Note the open hull positions of the MCG and the difference in size between the 5"/51 and 3"/50 in the superstructure mounts to the left. (USN-NHC)

Vought OS2U Kingfisher, Atlantic anti-submarine scheme, Medium Grey and White.

Vought OS2U Kingfisher, Semigloss Sea Blue, Intermediate Blue and White (scheme carried by Texas' Kingfishers after transfer to the Pacific in late 1944).

Texas, **December 1943**, Atlantic, Ms. 22.

California, **12 June 1944**, Saipan, Ms. 32/16d (starboard).

California, **12 June 1944**, Saipan, Ms. 32/16d (port).

Oklahoma, in January 1920, with range clocks and deflection scales, and flying-off platforms on B and X turrets, common after WW I. She was the last VTE-powered capital ship in the US Navy, but differed in no way externally from her turbine-powered sister. (USN-NARS)

After modernization, the Nevadas showed typical features, most noticeably tripod masts. Of particular interest in this view from the foredeck are the main battery rangefinder on the forebridge and Mk 33 AA directors on the bridge wings. Note the AA MGs on the foremast platform and mainmast top. This photo must date from just after Pearl Harbor because of the late alterations, Ms. 1 camouflage and obvious signs of the bridge fire. (Bob Cressman)

Nevada leads Oklahoma and one of the Pennsylvanias in maneuvers during the '20s. Roofs have been added to her spotting tops and the extensions of the flying-off platforms over the gun barrels have been removed as the growth in weight and performance of aircraft made these small platforms obsolete. The cloth 'sail' stretched between funnel and mainmast is a gunnery target. (USN-NHC)

Nevada was the only battleship in Pearl Harbor to get underway during the attack. Having survived the first wave with only minor damage, she attempted to break for open water. (Left) Coming under intense bomb attack off 10-10 Dock, she was soon taking on water forward and in great danger of sinking in the Entrance Channel. She was therefore beached. (Below) Note the Ms. 5 False Bow Wave standing out against the Dark Gray. (USN-NARS)

Extensively rebuilt, Nevada emerged from Puget Sound as a handsome fighting vessel with a modern, compact superstructure. This view dating from 1 July 1943 at San Francisco Bay, after the Aleutians campaign, just before her transfer to the Atlantic, shows the much strengthened AA fit. Equally important are the four Mk 37 AA directors with FD radar quartering the sky. Other radar includes two FC fire control sets, an SG on each mast and an SC on her mainmast. She is in an extremely faded Ms. 21. (USN-NARS)

Upon arriving in the Pacific, Nevada received a curvaceous Ms. 31a/6b. She is seen from Texas off Iwo Jima, 19 February 1945. (USN-NARS)

Entering Norfolk preparatory to transfer back to the Pacific, 17 September 1944, Nevada looks as she did supporting the D-Day and 'Dragoon' landings. She has either landed or never carried a quad 40 on the starboad side of X turret because there are now two single 20s at that location. She has an SK radar forward, otherwise she is identical to a year before. She is in the typical Atlantic Ms. 22. (USN-NARS)

Pennsylvania Class

BB38 Pennsylvania, BB39 Arizona

The **Pennsylvania** class represented a relatively modest advance over the preceding **Nevadas.** Length and displacement were somewhat increased and two further 14" guns were shipped, the main armament now being arranged in four triple turrets. Having finally been convinced of the superiority of turbines, two different designs were tried, Parsons in **Arizona** and Curtiss in **Pennsylvania.**

Conversion in 1929-31 altered the **Pennsylvanias** in a similar fashion as the preceding classes: tripods replaced cage masts, MCG was raised one deck, torpedo bulges were added and a catapult was fitted to the fantail. Later, a second catapult was fitted on X turret.

Both were at Pearl Harbor. **Arizona** fell victim to a devastating torpedo and bomb attack, being virtually broken in two at B turret. Salvage was never seriously considered. All tophamper was removed to the water level and she was left to serve as a memorial to those who died that day. **Pennsylvania** was in Drydock #1, suffering only minor damage. After hasty repairs, she rejoined the battleline in March 1942. In October, she re-entered Mare Island for a major rebuild. Twin 5"/38s replaced all previous MCG, the after tripod was replaced with a low tower and X catapult was removed.

Pennsylvania remained active in the Pacific throughout the war. She supported operations in the Aleutians, Gilberts, Marshalls, Palaus, Philippines (including action at Surigao Strait). She missed Iwo Jima and Okinawa but joined TF 95 for operations in the East China Sea in July 1945. While bombarding Wake in August 1945, she was torpedoed and seriously damaged.

Having been repaired only enough to make her seaworthy, she was a victim of both atomic tests and later served as an aircraft target. She finally sank on 10 February 1948, as much due to progressive flooding resulting from the torpedo hit off Wake as to her target duties.

The next class of battleships was strengthened by the simple expedient of making all four turrets triple mounts. One of the Pennsylvanias is seen at anchor in Hampton Roads, pre-modernization, 22 June 1920. (USAF)

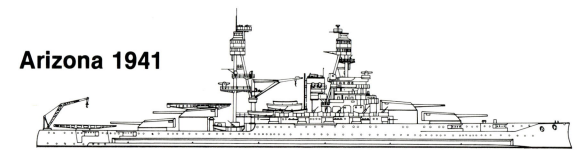

Arizona 1941

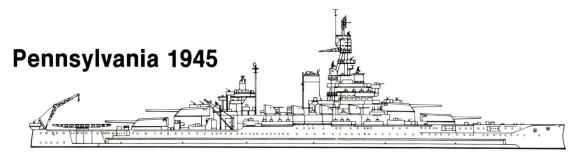

Pennsylvania 1945

After modernization, Arizona carries tripod masts, a catapult and resisted MCG. She is preparing to launch one of her spotting aircraft off the catapult. The reserve aircraft is alongside X turret. On this occasion she has an important guest, President Hoover, 1930. (USAF)

A typical MCG and AA suite for a US battleship at the outbreak of the war is visible in this dockyard view of Pennsylvania undergoing quick repairs at Mare Island, 26 February 1942. From bottom to top, the guns are: 5"/51s on the upper deck, 5"/25s in splinter-proof gunhouses and a quad 1.1" MG one deck up and single, shielded 20mm Oerlikons at the conning tower level. The 5"/51s were soon replaced as they could only be used against surface targets. The 5"/25s were handy, rapid-firing weapons but were replaced whenever possible by the newer 5"/38s as the VT proximity fuse made longer range heavy AA fire practical. The 1.1" was a failure that was replaced by quad 40mm Bofors as soon as possible. The 20 remained the standard light AA on US ships throughout the war, though its effectiveness was doubtful at best. (USN via Bob Cressman)

(Far Right) After her late 1942 rebuild, only the 20s remained from Pennsylvania's early fit. These appear in much greater numbers than before, at least nine are visible. 40s, canvas-covered, have replaced the 1.1s and one of her eight twin 5"/38 turrets is visible, having replaced both previous models of 5" gun. Two canvas-covered Mk 51 directors for her 40s can be see in the middle of this photo. (USN via Bob Cressman)

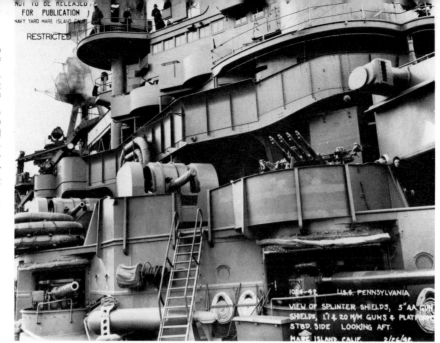

Early Anti-Aircraft Armament

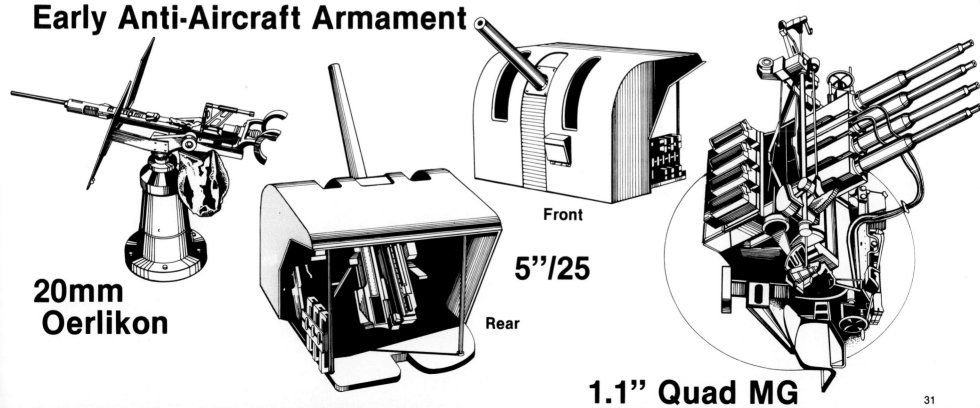

20mm Oerlikon

5"/25 — Front / Rear

1.1" Quad MG

While 'Battleship Row' was being raised, repaired and modernized, those battleships which weren't at Pearl Harbor or only received minor damage filled the gap. Two of these are seen at Adak, the forward base for TG 51.1, the support group for the reconquest of the Aleutians. Idaho is in the center, Pennsylvania to the right, surrounded by transports, escorts and a 'treaty cruiser', probably Louisville (CA 28). (USN-NARS)

(Above Right) After the Aleutian campaign, Pennsylvania underwent a modest, three month rebuild at which she lost her tripod mainmast and gained eight twin 5"/38 turrets. This dramatic view shows her firing on Guam, 21 July 1944. (USN-NARS)

Joining Oldendorf's TF 95 in the East China Sea, Pennsylvania is seen, 28 June 1945, in Ms. 21. She has a complete late-war radar suite. An SG, SK-2 dish and Mk 28 microwave main battery fire control radar top her foremast, while her mainmast carries an SP fine air search set. A Mk 13 fire control antenna tops her Mk 34 'cruiser'-type main battery director aft. Her two Mk 37 AA directors have the late Mk 12/22 radar, but her after AA directors are the new lighter type with microwave radar. (USN-NARS)

New Mexico Class

BB40 New Mexico, BB41 Mississippi, BB42 Idaho

In most features, the three **New Mexicos** were repeats of the preceding, successful **Pennsylvanias**. Length, displacement and armament were virtually identical, differing only in that most of the MCG were mounted a deck higher. The major recognition feature of this class was the 'clipper' bow, which characterized the remaining 'old' battleships. One other feature distinguished **New Mexico**, though not her sisters. She introduced the turbo-electric drive which also characterized the following designs. Each of **New Mexico's** turbines, rather than being geared directly to a propeller shaft, instead drove a generator which, in turn, drove huge electric motors connected to the four propeller shafts. The advantage of this system was in the greater flexibility it gave the engineering plant. Cruising speed could be maintained with some parts of the system cold or damaged and watertight sub-division was aided by the elimination of steam tubing, making for a more battleworthy vessel. The obvious disadvantage is to be found in the list of extra equipment required by the system, all of which took up hull space and added weight. The other unusual feature of this class was that it contained three rather than two ships, the usual number authorized by Congress in a Fiscal Year. **Idaho** was paid for by the sale of two old battleships (BB23 **Mississippi** and BB24 **Idaho**) to Greece.

Modernization in the early '30s altered the appearance of these ships considerably. As with all previous classes, the cage masts were removed. In this case, however, rather than tripods, the **New Mexicos** were fitted with a large tower foremast and smaller after superstructure inspired by the Royal Navy's **Nelsons**. All three received new boilers and geared turbines (**New Mexico** losing her turbo-electric drive), torpedo bulges and cranes for the previously added catapults.

New Mexico was at Casco Bay, **Idaho** and **Mississippi** were at Reykjavik when war broke out. All three were transferred to the Pacific immediately to shore up the defense of the West Coast, pausing only briefly at East Coast ports for quick refits at which time radar, light AA and splinter shields for the single, open 5"/25 AA mounts were added and all remaining 5"/51s were landed. Because these three were not at Pearl Harbor and suffered no serious damage throughout the war, they probably were altered less than any other class. Further minor refits gave each a differing heavy AA fit. **New Mexico** ended the war still mounting eight single 5"/25s, **Mississippi** carried 14 of the same, **Idaho** had 10 more modern 5"/38s in single gunhouses (identical to the main armament of most wartime destroyers).

After spending 1942 on guard duty off the West Coast, they then, singly or as a group, saw action in every Pacific campaign in which 'old' battleships participated. Having seen as much or more action as any other class, it was only fitting that all three should be at Sagami Wan to accept the surrender of the Japanese Empire.

New Mexico and **Idaho** were stricken in 1946. **Mississippi** continued to soldier on for many more years as a gunnery trainer and experimental vessel (redesignated EAG 128) in the place of **Wyoming**. Between 1946-47, all but one of her main turrets were removed and various 5"/38 turrets were mounted. In 1952, the then experimental Terrier SAM system was mounted aft, her last 14" turret being removed and a fully automatic 6"/47 twin turret was mounted forward. She was finally stricken from the Navy List in November 1956

Mississippi 1941

New Mexico 1945

Idaho 1945

With the New Mexico class, the 'clipper' bow that would characterize the last three classes, was introduced. This view taken 13 April 1919 show New Mexico with an aircraft on her B turret flying-off platform. The forward extension of that platform has been removed but the mounting brackets on the gun barrels remain. (USN-NHC)

Modernization left Mississippi with prominent towers fore and aft and two catapults. In this June 1941 view, she shows Mk 28 early-type AA directors on tall pedestals atop both towers with main battery directors one level lower. She has one of the first FC radars in between. She is in glistening Standard Navy Gray. (USN-NARS)

Idaho at Hvalfjord, Iceland, 2 October 1941, in Ms. 12. She still has a range clock on her forward tower. (USN-NARS)

New Mexico at Pearl Harbor, 4 December 1942, in a faded Ms. 21 (or this may be a last glimpse of Ms. 11). She now has a 'gallery' of 20s on each side between her funnel and after tower, plus numerous single mounts elsewhere. She has FC and SG radar forward and an SC on her mainmast. (USN-NARS)

Ten months later, definitely in Ms. 21, Mississippi looks almost identical. FC and SG fore and aft and an SK on her mainmast make up her radar suite. (USN via Floating Drydock)

Mississippi **received splinter camouflage in 1944, Ms. 32/6d. (Left) On her portside, the camouflage was composed of jagged bands of color, 21 July 1944. (USN-NARS) (Above) In contrast, her starboard scheme was curvy. More often than not, US camouflage schemes were symmetrical, but obviously not always. She is seen from** West Virginia, **20 July 1944. (USN-NARS)**

As with other classes, 1945 brought a resurrection of Ms. 21. Here, Idaho **is seen from** Nevada **off Iwo Jima, February 1945. Alone in this class,** Idaho **had her 5"/25s replaced by single 5"/38s in gunhouses. (USN-NARS)**

Off Okinawa, 11 April 1945, Mississippi has shed her Ms. 32 for Ms. 21. (Left) This aerial view shows how AA has taken over virtually every square foot of deck space between B and X turrets. (USN-NARS) (Below) She still retains her two old Mk 28 AA directors, now aided by microwave radar, and her open 5"/25 mounts. (USN-NARS)

(Above Right) Kamikazes posed a very real threat, particularly to those battleships without the latest AA or fire control. Fortunately, the 'old' battleships proved again and again to be tough customers. Mississippi was hit on one of her 5" mounts on 9 January 1945, but never left her post at Lingayen Gulf. Note the fusing and bearing information painted on the inside of the near 5"/25 guntub. (USN-NARS)

(Right Center & Below) Off Okinawa, New Mexico took a kamikaze on her funnel, 12 May 1945. An auxiliary is alongside helping fight the fire. She was under repair at Leyte until August, missing any further action but deservedly getting in on the surrender at Tokyo Bay with TG 32.9. (USN-NARS)

Tennessee Class

BB34 Tennessee, BB44 California

The **Tennessees** were virtual repeats of the **New Mexicos** with only a few significant differences. All now had turbo-electric propulsion. The most obvious external differences were a more built-up bridge structure, a reversion to a two-funnel arrangement and the total elimination of all hull casements making for a smoothly curved hull line. Refits in 1929-30 added catapults and heavy AA. Both were scheduled for a major modernization in 1939 but the worsening world situation caused those plans to be delayed and eventually abandoned.

Both were at Pearl Harbor on 7 December 1941. **Tennessee** was inboard of **West Virginia** which protected her from serious damage. The rapid repairs which followed in early 1942 included the replacement of the after cage mast with a stump tower and the addition of gunhouses and splinter shields for her 5"/25s. **California** didn't escape so lightly, being torpedoed and sunk in the Japanese attack. Only considerable effort by her crew kept her from capsizing. She was raised 25 March 1942 and drydocked. After her hull was made watertight, she sailed for Puget Sound, not rejoining the battleline until January 1944. **Tennessee** served briefly with TF 1 until September 1942, when she began a nine month modernization. Upon emergence, both ships were virtually identical. They carried rounded tower masts fore and aft with the single funnel faired into the foremast. A major torpedo bulge extending up into the superstructure amidships gave their hulls a distinctive notch just aft of the B turret. Eight twin 5"/38 turrets replaced the earlier heavy AA, all MCG being deleted.

Tennessee arrived back from the yard in time to join the Kiska operation. Thereafter, she participated in the Gilbert, Marshall and Kavieng bombardments. After **California** rejoined the fleet, they were, together or separately, at the Marianas, Palaus, Philippines (including Surigao Strait), Iwo Jima and Okinawa. Both were active in the East China Sea, raiding the Shanghai area. **Tennessee** was at Wake Island in August 1945 with **Pennsylvania**.

Having been extensively modernized, both **Tennessees** were placed in reserve and then mothballed rather than being immediately deleted. Both were finally stricken and scrapped in 1959.

The first of the 'Big Five', the turbo-electric battleships launched between WW I and the Washington Treaty, California **eases out of Mare Island, the largest warship constructed on the West Coast to that time. (USN-NARS)**

An attempt was made with the Tennessees **to establish a 'standard' bridge design for all future classes. This close-up shows** Tennessee **in April 1921. Note the 'flying bridge' connecting the tower bridge with the extremely tall conning tower. (USN-NARS)**

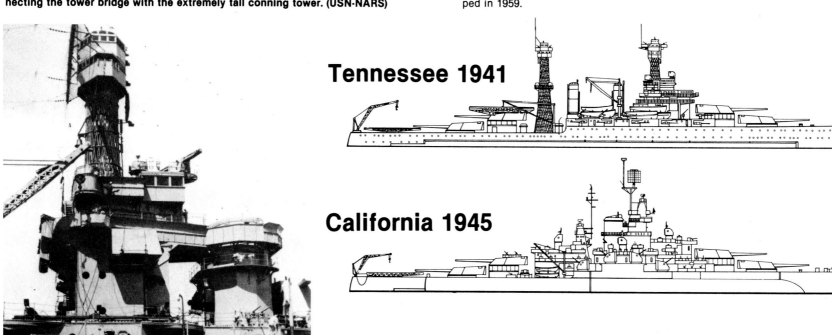

Tennessee 1941

California 1945

Both Tennessees were at Pearl Harbor on that Sunday, but suffered different fates. (Left) Calfornia received damage sufficient to cause her to sink, only her mooring ropes and harbor tugs keeping her from capsizing. In the background is the rest of 'Battleship Row' with Neosho in between. The latter survived this attack without a scratch only to be sunk at Coral Sea, five months later. (USN-NARS) (Right) Tennessee was inboard of West Virginia, being spared the torpedo damage that proved fatal to others. Her most serious damage came from burning oil leaking from Arizona immediately astern. In sinking, West Virginia jammed Tennessee against the forward quay which had to be blown in order to extricate her. Both ships are in Ms. 1. (USN-NHC)

Tennessee, because of her relatively minor damage, rejoined the fleet quickly. This May 1942 photo shows her not much changed from pre-war. She now has a short tower in the place of her mainmast, shields for her 5"/25s and an FC radar on her foretop. She is in the short-lived Ms. 11. Note the rudder striping on her catapult aircraft. (USN-NARS)

Having helped plug the gap in the immediate post-Pearl Harbor period, Tennessee was relieved in September 1942 for a nine-month major rebuild. She emerged totally changed with a wholly new superstructure and AA and a massive torpedo bulge. About all that remained unaltered was her main armament. She now shipped eight twin 5"/38 DP turrets plus 10 quad 40mm and multiple single 20mm. Each of four Mk 37 AA directors mounts an FD radar. She also has Mk 8 (FH) radars on her main battery directors, an SG on her foremast and SC-2 on her mainmast. She is in Ms. 21, May 1943. (USN-NARS)

(Above Left & Left) California took much longer to repair, not rejoining the fleet until the end of January 1944. When she did, she looked virtually identical to her sister, only search radar and camouflage differing. She mounted more quad 40s than Tennessee, requiring the addition of a 20mm gallery atop X turret to fit the necessary light AA. She wears Ms. 32/16d which looked similar, but was far from identical, port (12 June 1944) and starboard (28 October 1944). (USN-NARS)

Amid a mass of ships, part of TG 52.17 gathered for Operation 'Forager', Tennessee shows off her new Medium Pattern Scheme, Ms. 32/7d. Honolulu (CL48) is in the foreground, an escort carrier and another light cruiser are in the background, 12 June 1944. (USN-NARS)

Tennessee off Okinawa, March 1945, shows her reversion to Ms. 21. Marines in LVTs head toward the beach at the beginning of the bloodiest island campaign. The big guns of the 'old' battleships were decisive, particularly at Shuri in May. (USN-NARS)

At war's end, California carries the latest radars, Mk 12/21s on her AA directors, Mk 13s on her main battery directors and an SP fighter control/fire air search set on her mainmast, 29 September 1945. (USN-NARS)

Colorado Class

BB45 Colorado, BB46 Maryland, BB47 Washington (cancelled), BB48 West Virginia

With the exception of main armament, four twin 16"/45 turrets in the place of triple 14", and somewhat thicker armor, the **Colorados** were repeats of the **Tennessees**. This class of four was authorized in 1916 as the first part of a general construction program called for in the 1916 Navy Act. That same act authorized the six **South Dakota** class battleships and six **Lexington** class battlecruisers. Only three of those sixteen were ever to be completed as designed. Under the terms of the Washington Treaty, work on **Washington** was halted in 1922 when she was 76% complete. Like the **Tennessees**, these ships weren't seriously modified before WW II. They, too, were scheduled for modernization immediately before the war. Only **Colorado** actually entered a dockyard, Puget Sound in June 1941, before war changed all plans.

Because **Colorado** was undergoing refit, she missed damage at Pearl Harbor, but her two sisters were both there and were both hit. **Maryland** received only minor damage and was ready for action again in February 1942. **West Virginia** was the most seriously damaged of the battleships to be raised and repaired. She was the last to rejoin the fleet, not until July 1944.

Colorado's rebuild was cut short by the war. In March 1942, she emerged not much changed in appearance. A torpedo bulge (fitted to **Maryland** prewar), was the only visible sign that she had been in dock at all. **Maryland** emerged looking virtually identical. During the course of 1942, both returned to the dockyard for minor work. Their cage mainmast was shortened and 5"/25 AA mounts replaced by 5"/38s in shielded single mounts. A flying platform carrying six single 20mm mounts was fitted each side abreast the fore funnel.

Maryland and **Colorado** formed a two-ship subdivision that patrolled the Midway area, then Fiji-Noumea until late 1943. Both were at Tarawa in November and the Marshalls in January 1944, then both returned to Puget Sound for further rebuilding. At that time, the stump cage mainmast was replaced by a tower structure. Both were in action again in time for the Marianas operation.

West Virginia, when she at last emerged from repairs, looked nearly identical, except for main armament, to the rebuilt **Tennessees**. She returned in time to participate, along with **Maryland**, in the Palau operation. The same two were at Surigao Strait. All three were together for the first time off Leyte in November 1944. Thereafter, in varying combinations, they participated in all the remaining major Pacific operations. During 1945, **Colorado** had gunhouses mounted for her single 5"/38s and **Maryland** had her single mounts replaced by twin turrets. **Colorado** and **West Virginia** were in Tokyo Bay for the surrender.

All three went into reserve soon after war's end and were mothballed in 1947. They were deleted and scrapped in 1959.

With a new, heavier main battery, Maryland represented the peak of US battleship development before the Washington Treaty. With multiple awnings to protect the deck from the tropical sun, she negotiates the Panama Canal, c. 1930. (USN-NHC)

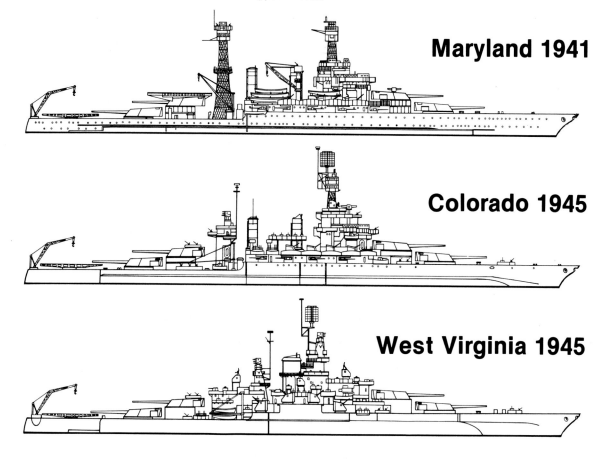

Maryland 1941

Colorado 1945

West Virginia 1945

West Virginia **was the most severly damaged battleship to be raised and repaired. Here she has been refloated and is being moved to Drydock #1, June 1942. (USN-NARS)**

Relatively undamaged at Pearl Harbor, Maryland **shows off her Ms. 1 camouflage. She was inboard of the capsized** Oklahoma. **The motor launch is from** Detroit **(CL8), 8 December 1941. (USN-NARS)**

For the trip back to Puget Sound, West Virginia's **turbo-electric plant was made operational and all unnecessary tophamper removed. Note the extensive patches on her portside. She would be rebuilding until mid-'44. (USN-NARS)**

Colorado **had just begun a major rebuild when the war broke out. Work in progress was hastily completed and she was back with the fleet by the end of March 1942. So hasty was that completion that she didn't have time to acquire the shields for her 5"/25s or stump tower aft like** Tennessee. **The only evidence of her months in the yard is the faintly visible torpedo bulge and FC radar on her foretop. (USN-NARS)**

Maryland **in 1944, looking virtually identical to** Tennessee **before the latter's big 1942-43 rebuild. Her two old AA directors have FD radar while there is an FC main battery set on each top, SGs on each mast and an SK forward. These views show the difference in the two sides of her camouflage scheme, starboard (Puget Sound, 25 April 1944) and port (off Saipan, 12 June 1944).** Louisville (CA28) **is in the background. (USN-NARS)**

Except for the difference in the paint job, Colorado was virtually indistinguishable from Maryland after their 1944 refits. (Above) Off Leyte, Colorado looks rather weatherbeaten. Her radars have been censored but presumably were similar to her sister's, 27 October 1944. (USN-NARS) (Right) A rare view of her other side, off Luzon, 6 January 1945. (USN-NARS)

Off Leyte on 24 October 1944, Maryland isn't on fire but using smoke pots fore and aft. Although faintly visible here, she still is in her splinter camouflage. That evening, she was in action at Surigao Straits, firing 48 salvos, ranging on West Virginia's splashes because her FC (also Pennsylvania's and Mississippi's) proved ineffective. (USN-NARS)

Colorado retained her splinter camouflage to the end. She plies her trade in this view, off Okinawa, 29 March 1945. At this point, she has microwave radar dishes on her AA directors and an FH fire control set aft. (USN-NARS)

West Virginia's AA battery opens up. This is probably a drill, note the absence of life jackets. Ready service rounds were stored around the inside of the 40mm guntubs, covered by canvas flaps when not needed. A Mk 51 director, controlling one or more 40mm mounts, is in its own small tub in the foreground. (USN-NARS)

(Above & Below) In July 1944, West Virginia finally left Puget Sound Navy Yard, rebuilt in a nearly identical fashion as the Tennessees, with a streamlined tower bridge and funnel structure, up-to-date radar fit and powerful AA. She is in Ms. 32/7d. (USN-NARS)

Not nearly as extensively rebuilt as West Virginia, but more altered than Colorado, Maryland shows the appearance of power and technology that marked the ultimate development of the 'old' battleship in WW II. The visible electronics are, from the stern: a Mk 13 atop the after main battery director, an SP fighter control radar with BO IFF on the mainmast, an IFF 'skipole' and TBS antenna on the port mainmast yardarm, Mk 12/22 radars on her Mk 37 AA directors, SG on her foremast, TDY-1a radome below and just aft and twin ECM domes on either side and below, SK air search on her foretop with microwave Mk 28 fire control set just in front and at each bridge wing, just below the Mk 37 directors, is a 'spoked wheel' ECM antenna, 5 August 1944. (USN-NARS)

Radars

No limit has historically been more restrictive on a battleship's ability to project its power than that of vision. The 'decisive' engagement of WW I, the battle of Jutland, was not decisive precisely because action was continually interrupted by poor visibility. Obviously, any new technology that could pierce smoke or dark would give the possessor a tremendous advantage. The US Navy came into possession of just such an advantage when in 1937 the first experimental ship-borne radar (Radio Detecting & Ranging) set was fitted on the destroyer **USS Leary.** This first Naval Research Laboratory set included a 10 ft. sq. antenna mounted on the barrel of one of **Leary's** 5" guns, operating on a frequency of 200mc. The power of this first set was limited, the greatest detection range achieved being 16nm. Greater power would be necessary before radar would become a useful tool. The invention of the ring oscillator in 1938 solved that problem to the point that a land-based test system was able to track a wooden aircraft at 100nm. In March 1938, development began on a sea-going version, XAF, which was installed in **New York** in December and satisfactorily tested in early 1939. 20 examples of a production variant, CXAM, were ordered from RCA, 19 of which had been installed onboard ships by 7 December 1941. A few, improved CXAM-1 were also produced. With higher power and limited tunability, CXAM-1 became SC, the first mass-produced radar, 500 of which were ordered from GE.

Before the end of the war, technology advanced from those early, often crude, sets to a myriad of refined, purpose-built radars that took possession of mastheads, supplanting the human eye. Between sea and air search radars with their associated IFF (Interrogation Friend or Foe) and fire control radars for all guns from 16" down to 40mm, fighter homing beacons, long-wave, short-wave and TBS (Talk Between Ships) antennas and jammers to foil enemy radars and ECM (Electronic Counter-Measures), the masts and yardarms of US battleships looked like electronic playgrounds, sprouting apparatus of various shapes and sizes. It was the superiority of Allied electronics to that of the enemy which, as much as anything else, assured the victory. The battle of Surigao Strait was lost by Nishimura from the instant he registered on a US radarscope.

Radar development followed two main avenues, search and fire control, given S and F designators by the US Navy. The first operational search radar, CXAM, was just being introduced to the fleet at the time of Pearl Harbor. **West Virginia** had CXAM-1 at that time, but it was not yet operational. The first operational fire control set, Mk 3 or FC, for main battery direction, had been installed on **Idaho** and possibly the other **New Mexicos** before Pearl Harbor. Thereafter, progress in both areas was rapid. Heavy AA directors soon mounted the Mk 4 or FD. The CXAM was replaced by the SC and then by increasingly advanced sets: SK, SC-2 and SK-2. Radars specifically dedicated to surface search, as opposed to air search, began to appear in 1942. The SG surface search set was standard on all large ships. Fire control sets likewise progressed rapidly, through the Mk 8 (FH) and Mk 13. For AA fire control, the FD was supplanted by the similar looking Mk 12, often in conjunction with the Mk 22 'orange peel'. Later in the war, microwave 'dishes', such as the Mk 19, were integrated with existing 40mm director positions. As deadly as the kamikaze attacks of 1944-45 proved to be, they would have been catastrophic had the electronic revolution not transformed gunnery from an art to a science.

The XAF was among the first radar sets and, with the similar CXZ, was the first fitted to a US battleship. The mounting here on New York **was experimental, January-March 1939. The XAF, developed by the Naval Research Laboratory, had a greater range than RCA's CXZ, fitted at the same time on** Texas, **which had better resolution. (USN-NHC)**

West Virginia **at Pearl Harbor mounted a CXAM-1 air search set, which, unfortunately, wasn't yet operational. This first limited production radar was supplanted by the SC during 1942. (USN-NHC)**

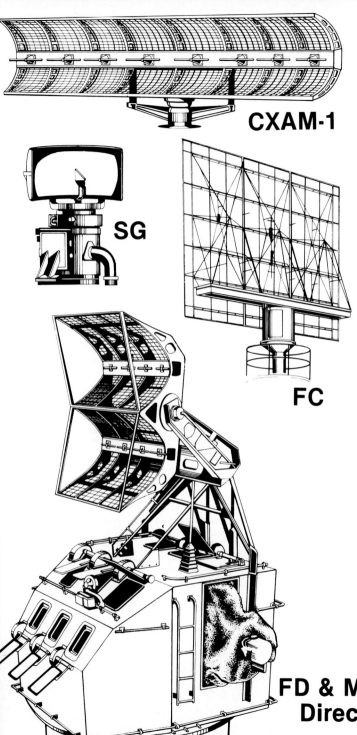

CXAM-1

SG

FC

FD & Mk 37 Director

Fire control radars became available at almost the same time as the first production air search sets. This view of *New Mexico's* foretop shows an FC (Mk 3) main battery set, just installed. At this time, an air search set hadn't yet been fitted. In the background is an Mk 28 AA director showing evidence of the 'splotches' of Ms. 12 Mod. which *New Mexico* carried briefly, 31 December 1941. (USN-NHC)

US ship yards kept excellent photographic records of the changes and additions made on naval vessels. For internal use, these photos were often tagged with radar designations to assist BuShips in keeping track of which ships carried what radars. While the untagged views were released soon after the war, these negs often stayed classified for years. This view shows *Tennessee's* superstructure at Puget Sound NY, 1 May 1943. The BK and BL-2 antennas were for IFF systems. (USN-NAC)

SS1036 F6F Hellcat in Action

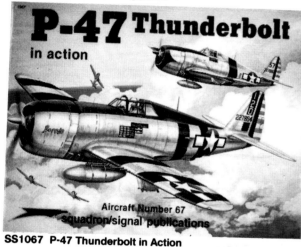

SS1067 P-47 Thunderbolt in Action

SS1039 Spitfire in Action

SS1045 P-51 Mustang in Action

squadron/signal publications presents

THE MAGNIFICENT SEVEN!

SS1044 Messerschmitt Bf109 in Action Part 1
SS1057 Messerschmitt Bf109 in Action Part 2

SS1029 F4U Corsair in Action

SS1059 A6M Zero in Action